D1282791

The Library of the Nine Planets™

JUPITER

Abby Young

The Rosen Publishing Group, Inc., New York

To John, with love and thanks

Published in 2005 by The Rosen Publishing Group, Inc.
29 East 21st Street, New York, NY 10010

Library of Congress Cataloging-in-Publication Data

Young, Abby.
Jupiter / by Abby Young. — 1st ed.
p. cm. — (The library of the nine planets)
Summary: Presents scientific discoveries about the atmosphere, moons, storms, and environment of this giant planet composed almost entirely of liquid and gas. Includes bibliographical references and index.
ISBN 1-4042-0168-8
1. Jupiter (Planet)—Juvenile literature. [1. Jupiter (Planet)]
I. Title. II. Series.
QB661 .Y68 2004
523.45—dc22

2003022334

Manufactured in the United States of America

On the cover: A photo of Jupiter's Great Red Spot taken by *Voyager 2.*

Contents

Introduction — 5

ONE — To Jupiter and Beyond — 7

TWO — King of the Planets — 14

THREE — Jupiter's Many Moons — 18

FOUR — A Hostile World — 25

FIVE — What's Next? — 33

Timeline of Exploration and Discovery — 38

Glossary — 39

For More Information — 41

For Further Reading — 43

Bibliography — 44

Index — 46

INTRODUCTION

If you look up into a clear night sky, you can see Jupiter. After Earth's moon, Venus, and sometimes Mars, it is the brightest object in the heavens. It looks like a very bright star but doesn't twinkle like a star. It has a bright, steady glow.

Because Jupiter can be seen from Earth, people have observed it for thousands of years. Ancient civilizations had no way of knowing that they were really looking at a planet, but they knew that it was different from the stars. They knew it was different because, unlike the stars, it seemed to move across the sky in a unique way. In fact, the word "planet" comes from the Greek word for "wanderer."

It was not until the invention of the telescope in the early 1600s that people were able to take a closer look at Jupiter. When the Italian astronomer Galileo Galilei turned his homemade telescope toward Jupiter in 1610, he saw a huge striped world with moons of its own. However, not much else was discovered about the giant mysterious planet for hundreds of years.

In the 1970s, the Pioneer spacecraft were sent to Jupiter. These were soon followed by the Voyager missions and eventually by *Galileo*, the spacecraft named after the astronomer, which was launched in 1989.

What has been learned about the planet from these missions is truly amazing. Jupiter isn't a planet as we normally think of one—it's actually a giant ball made completely, or almost completely, of liquid and gas. It has more moons than any other planet and is often viewed as its own mini solar system. The size of the planet is astounding, encompassing two-thirds of all the planetary mass in the solar system.

The environment of Jupiter is one of powerful storms, incredible lightning, deadly radiation, and winds many times faster than those of the strongest hurricanes on Earth. Scientists continue to be fascinated by the planet.

However, it is not just the planet that interests scientists and astronomers today. The moons, or satellites, of Jupiter are also subjects of intense study. And with sixty-three moons discovered so far, there is a lot to be learned.

Lately, much attention has been focused on the fact that liquid water has been detected on several of Jupiter's moons, suggesting that the moons may have or once have had some form of life since water is necessary for life as we know it. Europa, the sixth moon from Jupiter, is considered by many to be, along with Mars, among the likeliest places outside of Earth to find evidence of life.

All the planets in our solar system are fascinating, and each is unique. Let's take a look at what makes Jupiter different from the rest.

One

To Jupiter and Beyond

Ancient civilizations followed and charted the movement of Jupiter across the skies long before the invention of the telescope. When the telescope was invented in the early 1600s, people were able to get a closer look. It wasn't until the 1970s, however, that any detailed information was known about this giant planet.

The Pioneer Missions

In March 1972, the National Aeronautics and Space Administration (NASA) launched the spacecraft *Pioneer 10*. Its mission was to photograph Jupiter and its satellites and to gather information and data. This would be the first probe to attempt to pass through the asteroid belt, a dense path of asteroids between Mars and Jupiter. Many wondered whether it could be done. To everyone's great relief, *Pioneer 10* passed safely through the asteroid belt and arrived at Jupiter in December 1973.

    Pioneer 10 traveled to within 81,000 miles (130,354 kilometers) of the clouds at the top of Jupiter's atmosphere. It took the first close-up images of Jupiter and its moons before heading toward the outer solar system. It took measurements of the radiation belts, atmosphere, and magnetosphere, the surrounding region of space that is dominated by the object's magnetic field. It continued making important scientific

Pioneer 10, shown here being launched, was the first spacecraft to fly through the asteroid belt. It was also the first spacecraft to make direct observations and take close-up images of Jupiter. As of this writing, *Pioneer 10* had traveled nearly 8 billion miles from Earth.

observations in the outer solar system until its official mission ended in 1997.

Even after its mission ended, scientists continued to track the probe. *Pioneer 10*'s last signal was received in January 2003. Even though its power source is gone, the probe will continue to travel silently through space. In 2 million years it will pass near the first object in its path: the star Aldebaran in the constellation Taurus.

Pioneer 11 was launched in April 1973. Its mission was to investigate Jupiter, Saturn, and the outer solar system. It passed closer to Jupiter than *Pioneer 10*, coming within 27,000 miles (43,452 km) of the cloud tops, and sent back more photos of Jupiter and its moons. It also gathered information about the planet and its satellites, its magnetosphere, and solar wind. It then used the gravity of Jupiter to adjust its course and head to Saturn.

The last transmission from *Pioneer 11* was received in 1995 and communication is no longer possible. We do know, however, that the first object it will come close to is a star in the constellation Aquila. The trip will take 4 million years!

In addition to the numerous photos and information that the Pioneer spacecraft were able to obtain, they showed that travel through the asteroid belt was possible. This paved the way for a long line of spacecraft to be sent out to investigate and explore the universe.

The Voyager Missions

In August 1977, *Voyager 2* was launched. Sixteen days later, *Voyager 1* was launched. However, because of the different paths the probes would take, *Voyager 1* would reach Jupiter first. The mission objectives of these probes were to investigate Jupiter's atmosphere and the magnetosphere. They were also scheduled to analyze the satellites of Jupiter and Saturn, to study the ring system of Saturn, and to search for a possible ring system around Jupiter.

Voyager 1 reached Jupiter on March 5, 1979, and discovered that Jupiter does indeed have rings. *Voyager 2* was quickly reprogrammed to take a closer look when it arrived at Jupiter on

July 9, 1979. Both probes sent back stunning photos and scientific data, and continued on to Saturn and to the outer solar system.

Voyager 2 completed what is called the Grand Tour, visiting Jupiter, Saturn, Uranus, and Neptune. So far, it is the only spacecraft to have visited Uranus and Neptune.

The Voyager probes are still sending back information, and their missions to explore space outside the solar system are still in progress. *Voyager 1* is now the farthest away of all man-made objects. On November 5, 2003, it reached a distance of 8.4 billion miles (13.5 billion km) from the Sun. Both spacecraft have enough power to continue sending information until approximately 2020.

Ulysses and Cassini

Not all of our information and photos of the Jovian system (Jupiter and its moons) came from missions focusing on Jupiter. Sometimes information is gathered from missions to other planets or celestial bodies.

Ulysses is a joint project between NASA and the European Space Agency (ESA), an organization of fifteen European countries devoted to space exploration. Launched in October 1990

Message in a Spacecraft

Each of the Pioneer and Voyager spacecraft is carrying a message in case any type of intelligent life ever intercepts it. The Pioneer spacecraft are both carrying a plaque with pictures of Earth and earthlings. The Voyager spacecraft are both carrying recorded messages, which include greetings in more than fifty-five languages, sounds and music from our planet, and a message from President Jimmy Carter. Who knows when one of them could be found!

Voyager 1 was launched in the summer of 1977 along with *Voyager 2* to explore Jupiter's atmosphere, magnetosphere, moons, and a possible ring system. Shown here on August 27, 1977, *Voyager 1* is being prepared for its launch. At the time of this writing, *Voyager 1* is the farthest spacecraft in the universe. It reached 8.4 billion miles (13.5 billion km) from the Sun on November 5, 2003.

from the space shuttle *Discovery*, the spacecraft's mission was to study the polar regions of the Sun. However, in order to reach the necessary orbit, it needed to get a gravity assist from Jupiter. (A gravity assist is when a spacecraft falls into a planet's orbit and

Shown here is the *Cassini* spacecraft in NASA's Jet Propulsion Laboratory assembly room in Pasadena, California. Standing two stories tall and weighing nearly 11,670 pounds (5,293 kg), *Cassini* carries many high-tech sensors designed to conduct twenty-seven different tests to further our understanding of Saturn.

uses the planet's gravity to speed the spacecraft up—sort of like a gravity slingshot.) *Ulysses* passed Jupiter in February 1992 and sent back remarkable information about its magnetosphere, radiation, and Io, one of its moons.

Cassini is another example of one of these missions. *Cassini* is a joint project of NASA, the ESA, and the Italian Space Agency, and it was launched in October 1997 to Saturn. On its voyage to Saturn, it also needed a gravity assist from Jupiter. It flew by Jupiter in December 2000. It sent back about 26,000 images over a six-month period and provided a wealth of information about the planet, its rings, and its moons. *Cassini* will arrive at Saturn in 2004.

Galileo

Galileo was one of NASA's most ambitious projects. It actually consisted of two parts, an orbiter and an atmospheric probe. It was launched in October 1989 from the space shuttle *Atlantis* and began its long journey to Jupiter. Its flight path required a gravity assist from Venus and two from Earth to gain speed. The spacecraft was able to make detailed observations about these planets, the moon, and the asteroid belt.

In July 1994, *Galileo* was also able to witness one of the most spectacular shows in the solar system. Fragments of the comet

Shoemaker-Levy 9, which had broken apart when it flew close to Jupiter in 1992, crashed into the planet. This was the first time that anyone witnessed the collision of two solar system bodies. The comet fragments hit the planet with the force of thousands of nuclear bombs in an amazing show that lasted for days.

Galileo reached Jupiter in December 1995 and began its two-year mission of orbiting the planet (its mission was later extended several times). The next phase of the mission was about to begin. The atmospheric probe entered Jupiter's atmosphere on December 7, 1995, at 106,000 miles per hour (170,590 km/h), or 29 miles per second (46.7 km/s). Although the probe transmitted information for just under an hour and was eventually crushed by the incredible atmospheric pressure of the planet, it was able to send back remarkable data.

With *Galileo*'s propellant almost depleted, the spacecraft would soon have been impossible to control. It was feared that the spacecraft could eventually collide with Jupiter's moon Europa, so NASA opted to destroy *Galileo* by flying it into the planet on September 21, 2003, rather than risk contaminating the moon with any microbes that might possibly exist on the spacecraft. Europa is considered one of the likeliest places in the solar system outside of Earth to find evidence of life, and any contamination of the moon could pose problems for future studies.

Two

King of the Planets

Jupiter is a very different type of planet from Earth. It doesn't have mountains or craters. It doesn't have volcanoes or valleys. In fact, it doesn't even have a surface. It is one of the gas giants, which are planets composed entirely, or almost entirely, of gas and liquid. Saturn, Uranus, and Neptune are the other gas giants. The inner planets—Mercury, Venus, Earth, and Mars—are the terrestrial planets, which are the planets with solid, rocky surfaces. Pluto, being a small planet of rock and ice, does not quite fit into either of these categories.

An Appropriate Name

When the ancient Romans named Jupiter after the Roman king of the gods, they didn't know how appropriate the name was. Jupiter's incredible size makes it stand out from all the other planets. It is so big that if it were hollow, more than 1,300 Earths could fit inside it. In fact, all the other eight planets in the solar system could easily fit inside Jupiter, and there would still be room. The only bigger body in the solar system is the Sun.

Jupiter's diameter, or distance from end to end through the equator is 89,000 miles (143,000 km). In comparison, Earth's diameter is 7,926 miles (12,756 km), so Jupiter is about 11.2 times wider than Earth.

Jupiter's mass, or amount of matter, is 318 times greater than that of Earth. You might expect that a planet with such a

In this painting by Francois Verdier (1651–1730), Jupiter is shown at the top center. In Roman mythology, Jupiter is the supreme leader of the universe. This is a fitting name for the planet Jupiter. By far the largest body in the solar system, Jupiter is often called the king of the planets.

large mass would be quite dense, too, but that is not the case. Density is how tightly packed matter is. Jupiter has a density slightly greater than that of water, which is not dense at all. Earth is much smaller, but it is actually four times denser than Jupiter. Jupiter has a great deal of space for its mass to exist in, making it less dense.

One might think that such a huge planet would rotate slowly when compared with smaller planets, but Jupiter actually rotates faster than any other planet in the solar system. It completes one rotation in just less than ten hours (Earth takes twenty-four hours to complete one rotation). This incredible speed causes Jupiter to bulge out at the equator and flatten slightly at the poles, creating a shape called an oblate spheroid.

The two bright objects at the bottom right just above the horizon are Venus and Jupiter. Jupiter, the dimmer of the two planets, appears below and to the left of Venus, the brightest object in the image. When the two planets are positioned approximately one above the other from the viewpoint of Earth, as they are here, they are said to be in "conjunction."

A Very Distant Planet

Jupiter is the fifth planet from the Sun, and its orbit lies between those of Mars and Saturn. It orbits the Sun at an average distance of 484 million miles (779 million km). It is slightly more than five times farther from the Sun than Earth is. The distance is so great that it takes almost twelve years for Jupiter to orbit the Sun.

The closest Jupiter comes to Earth is about 366 million miles (589 million km). To give you an idea of how far that is, if there were a highway in space between Earth and Jupiter and you drove a car at a speed of 65 miles per hour (105 km/h), it would take 642 years to reach Jupiter. Even if you could travel at the speed of sound (761 mph [1,225 km/h] at sea level), it would take you 55 years! And that is at the closest point. When Jupiter is at its farthest, it is 602 million miles (968 million km) from Earth.

A Strong Attraction

Gravity is what keeps planets revolving around the Sun and stops them from spinning off into space. It keeps the Moon in orbit around Earth. It keeps us from flying off the planet. All bodies in the solar system have and are affected by gravity.

The more massive an object is, the stronger its gravitational pull. Because of Jupiter's large size and mass, it has a very strong gravitational pull. Only the gravitational pull of the Sun is stronger. This is one of the reasons that Jupiter has so many moons. Many of the smaller outer moons are believed to be asteroids that got caught in Jupiter's orbit by its incredibly strong gravity. Jupiter has also caught two other groups of asteroids. These are called the Trojans. They travel in the same orbit as Jupiter, one group just ahead and one group just behind Jupiter.

If you could actually stand on the surface of Jupiter—that is, if there were a surface to stand on—it would be quite difficult to move around. The gravity is slightly more than two and a half times stronger than it is here on Earth. That means that if you weighed 100 pounds (45 kilograms) on Earth, you would weigh 254 pounds (115 kg) on Jupiter. If you wanted to leave Jupiter in a spacecraft, you would have to travel faster than 133,000 miles per hour (214,043 km/h) to overcome its powerful gravity and escape its pull. That's about 37 miles (60 km) per second. At that speed, you could travel from New York City to Los Angeles, California, in a little more than one minute!

Three

Jupiter's Many Moons

Not only does Jupiter have the largest number of moons by far of all the planets in the solar system, but it also has some of the most fascinating moons. Among Jupiter's moons is the largest moon in the solar system, the most volcanically active body in the solar system, and some of the most promising places to find life in our solar system. Many of these moons are so large that they would be classified as planets if they orbited the Sun. Jupiter also has some of the smallest moons and has helped spark discussion on whether a new system of moon classification is needed.

The Galilean Moons

The four largest and most studied moons of Jupiter are called the Galilean moons, named after Galileo, who discovered them in 1610. Although we have known and observed the Galilean moons for almost 400 years, it was not until the Voyager missions in the late 1970s that we became aware of how fascinating and vastly diverse they are.

Ganymede

Ganymede is the largest moon in the solar system. It is the seventh moon from Jupiter and orbits at an average distance of 664,870 miles (1,070,000 km). It has a diameter of 3,268 miles (5,259 km), making it larger than Mercury and Pluto.

It's about three-quarters the size of Mars.

Half of the moon's surface is heavily cratered dark terrain, which is very old. The other regions are younger and consist of grooves and ridges, which have been created by movement of the moon's crust over time. It is most likely made up of a rocky core with a water-and-ice mantle and a crust of rock and ice. Recent evidence from *Galileo* suggests that an ocean of liquid water may exist below the icy crust.

Galileo made another interesting discovery about Ganymede. It is believed to be the first moon with its own magnetosphere, making it the first known example of a magnetosphere within another magnetosphere (Jupiter's) in the solar system.

Jupiter's moon Ganymede, shown here, has several different types of terrain. These different types are shown in different colors, which have been manually enhanced. The bright areas are the younger regions. The dark areas are older. The violet colors at the poles are believed to be from small particles of frost, which scatter the light.

Callisto

Callisto is Jupiter's second largest moon. Its diameter is 2,983 miles (4,801 km), making it about the size of the planet Mercury. It is the eighth moon from Jupiter and orbits at an average distance of 1,170,000 miles (1,883,000 km). The moon is believed to consist of a rocky core surrounded by water and ice. A thin atmosphere of carbon dioxide has been detected at its surface.

Callisto is the most heavily cratered satellite in the solar system, and its crust dates back 4 billion years. The largest craters have been partially erased by the flow of the icy crust over time.

Its largest impact crater is Valhalla, whose central region measures 373 miles (600 km) in diameter. The rings of the crater extend to 1,864 miles (3,000 km) in diameter.

Until recently, it was believed that Callisto was just another boring moon of rock and ice. However, information sent back from *Galileo* indicates that there might be an ocean of salty water beneath its surface. If so, could there be life there, or could life have once existed there?

Io

Io is the fifth moon from Jupiter and orbits the planet at an average distance of 261,970 miles (421,600 km). Its diameter is 2,256 miles (3,631 km), making it a bit larger than Earth's Moon. It is

Naming Jupiter's Moons

The names of the Galilean moons come from names of mythological companions of Jupiter, the Roman king of the gods. The moons were named by Simon Marius (1573–1624), who claimed to have discovered the moons at about the same time as Galileo. Galileo, however, publicized his findings first, so Marius's claims were impossible to prove.

Marius was allowed to name the moons and decided to base the names on a system suggested by German astronomer Johannes Kepler (1571–1630). This system was also used in naming moons discovered later. However, with a rapidly rising number of moons, names became harder to come up with. The pool of possible names has recently been expanded to include children and other relatives of Jupiter and Jupiter's companions.

Many of Jupiter's moons have not been named yet and only have numerical designations. Eventually, the names will have to be approved by the International Astronomical Union (IAU), which is the authority for the naming of all celestial bodies and any features on them.

Shown here is an image of volcanic activity on Jupiter's moon Io, taken by *Galileo* on February 22, 2000. The white and orange area to the far left is newly erupted hot lava. The two small white spots are eruption sites. The large orange and yellow ribbon is a cooling lava flow that is more than 37 miles (60 km) long. The dark L shape at the left is a lava flow from a previous eruption that took place in November 1999.

believed that the moon consists mostly of molten sulfur and its compounds or silicate rock. Recent data from *Galileo* suggests that Io has an iron core. There are few impact craters on Io, as its volcanic activity is constantly resurfacing it.

Io is the most volcanically active body in the solar system. Volcanic plumes of rock and sulfur rise 190 miles (300 km) above the surface. The intense volcanic activity is caused by the tremendous tidal forces, the gravity of other bodies, acting on Io, the friction of which keeps the subsurface crust in a liquid state. These tidal forces are so strong they pull on the moon's surface, causing it to bulge as much as 330 feet (100 m).

Europa

Europa is the sixth moon from Jupiter and orbits at an average distance of 416,880 miles (670,903 km). Its diameter is 1,951 miles (3,140 km), making it slightly smaller than Earth's Moon. It is most likely composed of silicate rock.

Europa is the smoothest body in the solar system. There are very few impact craters on the surface. The surface is all ice, and it is believed that craters have been filled in by water from below, which then freezes. Why doesn't all the water freeze? Europa is affected by the strong gravity of Jupiter and neighboring moons. The flexing of the planet from the tidal forces it is subjected to causes friction, which melts the ice.

From this resurfacing scientists see that Europa has two of the basic necessities for life as we know it: water and heat. Add to that some organic matter possibly contributed by an asteroid or comet, and the building blocks for life are there. Because of the possibility of some life-form existing, or having existed, on Europa, the moon is a high-priority exploration destination for NASA. There are several missions to Europa being looked into at this time.

Other Moons

Jupiter's other moons are much smaller. The next two in size after the Galilean moons are Himalia, which is 115 miles (185 km)

Tidal Forces

The tidal forces on Jupiter's moons can be compared to the rise and fall of ocean tides on Earth. The oceans rise and fall because of the gravitational pull of the Moon and Sun. Jupiter's moons are affected by Jupiter's incredible gravitational pull and by the gravity of other moons. This causes a tug-of-war effect, which, if strong enough, can actually cause the surfaces of the moons to bend up and down.

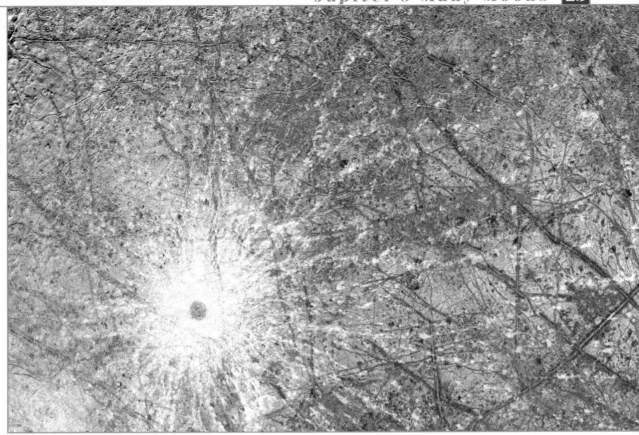

Shown here is the young 16-mile (26-km) impact crater Pwyll on Jupiter's moon Europa. Pwyll is thought to be one of the youngest features on Europa's surface. The bright white lines reaching out from the center are believed to be made of fresh ice particles. These lines are over 621 miles (1,000 km) long and extend over many different types of terrain, which indicates that they are new.

across, and Amalthea, which is 106 miles (170 km) across. The rest are even smaller. Some of the outer moons move in the opposite direction of Jupiter's rotation. This suggests that they are captured asteroids.

The number of newly discovered moons is rising quickly. Since 2000, forty-six new moons have been discovered. Although most of the previously discovered moons were first seen from images taken by spacecraft, this latest and largest group was discovered from Earth. Dr. David Jewitt and his colleagues at the Institute for Astronomy at the University of Hawaii have discovered dozens in the last few years. The moon count at the time of this writing stands at sixty-three.

With these discoveries, the number of known moons in the solar system has risen past 120. With *Cassini* scheduled to arrive at Saturn in 2004, the number is expected to continue to rise rapidly. Powerful new telescopes and new technology will cause this number to rise even more. Some even expect it to double in the next few years and possibly someday triple.

In early 2003, the smallest known satellite was discovered orbiting Jupiter. It is only 0.62 miles (1 km) in diameter. Then, in early 2004, two new objects measuring only 1.2 miles (2 km) in diameter were confirmed to be moons of Jupiter. With smaller and smaller moons being discovered at such a rapid rate, some people are wondering if there will need to be a new definition of "moon" or a new classification system to differentiate between different types and sizes of moons. If a satellite were discovered that was only a few feet across and shaped like a potato, would it be considered a moon? What about something the size of a pebble? By applying the current definition of a moon—anything that orbits a planet—they would. David Jewitt, however, is not interested in redefining "moon" or setting any limits. As he was quoted on Space.com, "Is a small dog not a dog because it is small?"

Four

A Hostile World

Below the colorful bands of clouds in Jupiter's upper atmosphere lies a very hostile environment. Powerful storm systems, winds much stronger than those of the strongest hurricanes on Earth, powerful lightning, and deadly radiation are all a part of what makes Jupiter so fascinating and interesting to astronomers and scientists. But this environment also makes Jupiter a very difficult planet to explore.

Much of the information we have about the atmosphere of Jupiter was gathered from the *Galileo* probe in 1995, which lasted just under an hour before being destroyed by the atmospheric pressure of the planet. But even in such a short time, the probe was able to transmit a lot of information and give us a better idea of what lies below the clouds.

Below the Clouds

Jupiter's atmosphere is 99 percent hydrogen and helium. The remaining 1 percent contains methane, ammonia, water vapor, and trace elements. Below the clouds is a sea of liquid hydrogen and helium 13,000 miles (21,000 km) deep. As you go farther into the interior of the planet, the pressure increases drastically, and the hydrogen turns into a liquid metal, similar to the mercury in a thermometer. This liquid metallic hydrogen could be as deep as 25,000 miles (40,000 km). Liquid metallic hydrogen is a good conductor of electricity, and it is this layer that is responsible for Jupiter's strong magnetic field.

At the center of the liquid metallic hydrogen is the core of the planet. The temperature there may be as high as 55,000° Fahrenheit (30,000° Celsius), and the pressure is millions of times stronger than the atmospheric pressure at Earth's surface. Some believe that Jupiter has a rocky core, while others believe the core is not solid but an extremely dense liquid. The core is estimated to be one and a half times the size of Earth's diameter.

The heat from the core is generated outward through a process called convection. It is this heat that powers the storms on Jupiter. On Earth, it is the heat from the Sun that drives the weather patterns. On Jupiter, the weather patterns are caused by heat and energy

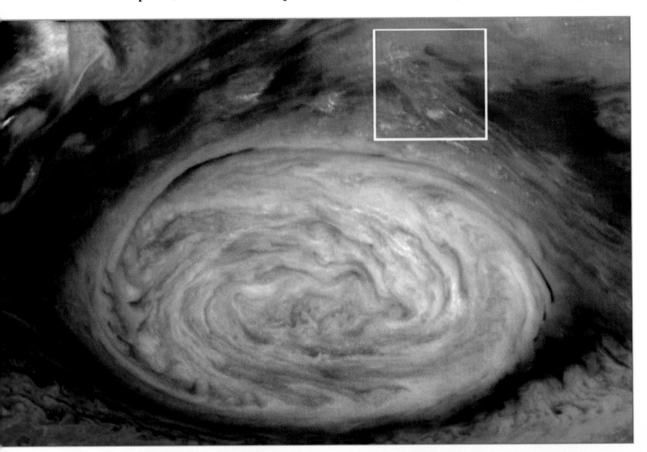

This storm on the outer edge of Jupiter's Great Red Spot was taken by *Galileo* on June 26, 1996. The white cloud is 620 miles (1,000 km) across. It is believed that this cloud is made of water and is similar to those on Earth. It is unknown whether this storm is producing rain or snow, but it is believed that this storm is producing lightning.

from within. Jupiter actually gives off more heat and energy than it receives from the Sun.

A Revised Theory

Through a telescope, Jupiter appears striped. These bands of color are caused by the different cloud layers in the upper atmosphere. The different colors are caused mainly by the chemical composition and temperature of the various layers of clouds. The clouds are in constant motion, powered by winds as strong as 400 miles per hour (644 km/h). In comparison, the most powerful hurricanes on Earth have winds between 150 and 200 miles per hour (between 241 and 322 km/h). The temperature at the cloud tops is a chilly -243°F (-153°C).

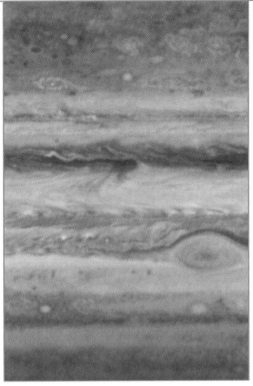

Shown here are the different types of clouds on Jupiter. The bands, caused by the different cloud layers in the upper atmosphere, are moving at extreme speeds. Depending on the color of the band, scientists can tell the altitude, composition, and temperature of the clouds.

The stripes, or bands, are referred to as belts and zones. The darker bands are called belts, and the lighter bands are called zones. For a long time, it was believed that the zones were areas of rising atmosphere and that the belts were areas of descending atmosphere.

Recent photos by *Cassini* have changed that theory. Images of individual storm systems rising in the darker areas have led astronomers to believe that the belts are areas of rising air and that the zones are areas of descending air. Dr. Tony Del Genio, an atmospheric scientist at NASA's Goddard Institute for Space Studies, described this new finding in a 2003 Jet Propulsion Lab press release as "the opposite of expectations for the past 50 years."

The Great Red Spot and the Great Dark Spot

Thousands of storms the size of the largest storms on Earth rage on Jupiter. The planet's largest and most well-known storm system is the Great Red Spot. First identified by English scientist Robert Hooke in 1664, it is a powerful storm, much like a hurricane on Earth, that has raged for hundreds of years. It is so big that at least two Earths could fit inside it. The size and shape of the storm may change slightly over time, but overall its position on the planet remains constant.

The Great Red Spot was thought to be the biggest feature on Jupiter, but photos taken by *Cassini* in 2000 show another feature just as large. Referred to as the Great Dark Spot, it is a dark cloud two times the size of Earth that appears around the planet's north pole. It was first glimpsed in an image taken by the Hubble Space Telescope in 1997, but it seems to appear and disappear over a period of months. Although astronomers are not sure what causes it to appear, some believe that it is a side effect of auroras on Jupiter. An aurora is a phenomenon caused when charged particles, often from solar wind, strike gases in the atmosphere, causing them to glow.

One Long, Cold Season

Seasons on Earth are not caused by the distance between Earth and the Sun. Rather, they are caused by Earth's tilt. Earth's 23.5-degree tilt means that the Sun shines directly over the Northern Hemisphere during part of the year, and shines directly over the Southern Hemisphere during the other part of the year. Jupiter, however, has a tilt of only 3 degrees. Because of this, the temperature does not vary much throughout the year, staying at about -243°F (-153°C) at the cloud tops.

This image of Jupiter's Great Red Spot was taken by *Voyager 1* in March 1979. The distance in this image from top to bottom is a gigantic 15,000 miles (24,000 km). The white oval in the bottom right corner is a feature of the storm that formed around the year 1960 and is still raging today.

The Magnetosphere and Radiation Belts

The region of space surrounding a celestial body that is dominated by the object's magnetic field is called its magnetosphere. Jupiter's magnetosphere isn't round like the planet, though. Solar wind pushes on it, forcing it into more of a teardrop shape. Jupiter's magnetosphere is so large that, if it were visible, Jupiter would appear as large as the Moon in Earth's sky.

Charged particles trapped within the magnetosphere form powerful radiation belts in its inner portions. Recent data from *Cassini* shows that these radiation belts are more powerful than previously estimated. It is believed that this is the most deadly radiation area around any of the planets. Not only is this dangerous radiation a concern for spacecraft, it would be deadly for a human being.

Jupiter's Rings

Saturn's ring system has been known about for hundreds of years, but it wasn't until *Voyager 1*'s discovery in 1979 that we learned of Jupiter's ring system. It is now known that all the gas giants have ring systems.

Recent images have shown that Jupiter's ring system is made up of three main parts: the halo, the main ring, and two faint gossamer

Star Quality

In many ways, Jupiter resembles a star more than a planet as we think of one. In fact, if Jupiter were between 50 and 100 times larger, it would have become a star. As it is, the planet is too small and its core is too cool to become a star.

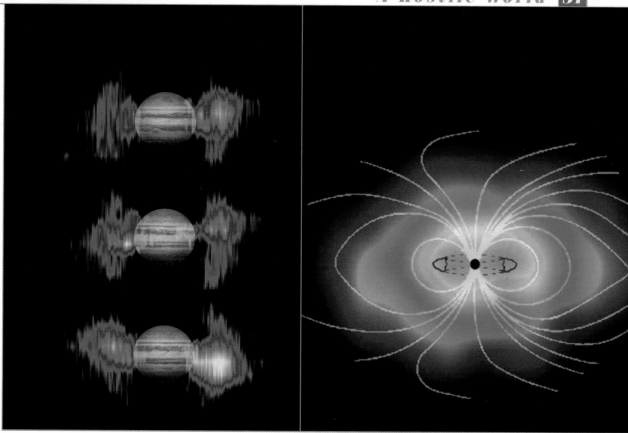

On the left are three images of Jupiter's radiation belts taken over a ten-hour period. On the right is an image of Jupiter's magnetosphere. To give a sense of its size, the black circle in the middle is Jupiter. Jupiter's magnetosphere is the largest feature in the solar system. If it were visible to the human eye, it would appear about the size of the moon as viewed from Earth. Both images were taken by *Cassini.*

rings. Recent data from *Galileo* has shown that the ring system is made up of tiny dust-sized particles created from meteor impacts into the small inner moons. These moons—Metis, Adrastea, Amalthea, and Thebe—are often called the ring moons.

The ring system begins at about 62,000 miles (100,000 km) from the center of Jupiter and extends to the orbit of Thebe, approximately 137,880 miles (221,900 km) from Jupiter. Unlike the rings of Saturn, Jupiter's rings do not appear to contain ice.

The rings are not the only things that encircle Jupiter. There are also giant doughnut-shaped gas clouds surrounding the planet. One is known as the Io plasma torus. Jupiter's magnetosphere actually strips off one ton (1,000 kg) of material from Io's surface every second. The material forms this doughnut-shaped cloud, called a torus, following Io's orbit. Another gas cloud was recently discovered surrounding Europa. It is believed that ion radiation from Jupiter damages Europa's surface, pulling apart water molecules and sending them out into Europa's orbit.

Five

What's Next?

What does the future hold for our exploration and attempts to understand Jupiter and its moons? Would we ever be able to send a man or woman to Jupiter? At the moment, this appears unlikely. The time it takes to fly to Jupiter alone makes the prospect of manned spaceflight nearly impossible. That, combined with the hostile environment and deadly radiation of the planet, makes the task unachievable at this time. Perhaps in the future, technology will be advanced enough to make it possible.

Even though a manned mission would be impossible, what about another unmanned spacecraft? Scientists are currently working on new ways to investigate the planet. It has been suggested that the best way to further observe Jupiter and its atmosphere would be a new and different type of probe. Perhaps the best type of probe would be some sort of balloon-like craft, one that could float above the atmosphere, thereby avoiding the crushing atmospheric pressure.

At the moment, there are no scheduled future missions to Jupiter. However, there are proposed missions to investigate some of its moons.

Probes to Europa

Europa is believed by many to be one of the most likely places to find evidence of life in the solar system. According to Ron Greeley, chair of the Europa Focus Group, who was quoted on

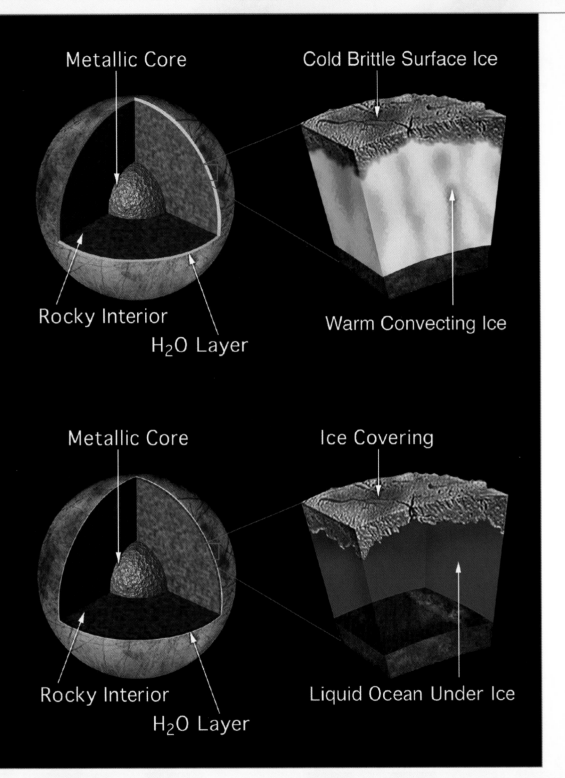

Metallic Core

Cold Brittle Surface Ice

Rocky Interior

H_2O Layer

Warm Convecting Ice

Metallic Core

Ice Covering

Rocky Interior

H_2O Layer

Liquid Ocean Under Ice

Shown here are two proposed models of what possibly lies beneath the surface of Jupiter's moon Europa. The top model shows warm ice below a cold, brittle surface. The bottom model shows a layer of ice 10 miles (16 km) thick with an ocean below it, which reaches to a depth of 60 miles (97 km). If this ocean does exist, it would be ten times deeper than any ocean on Earth.

the NASA Astrobiology Institute Web site, "If we think about the basic ingredients for life, we need liquid water, we need an energy source, and we need the right organic compounds. Europa seems to be a place where those three ingredients can be found." The NASA Astrobiology Institute's Europa Focus Group was founded in 2001 and brings together scientists of different disciplines to study the prospect of life on Europa.

Europa is considered a high-priority exploration destination, and a number of future missions to the moon are currently being investigated. One of the proposed missions is the Europa Ice Clipper, which would attempt to obtain samples of the surface. There is also the Europa Ocean Explorer, which would land on Europa and explore what is beneath its icy crust with the help of cryobots (long, narrow probes that can melt ice and can obtain data from the water below) and hydrobots (remote-controlled submarine-type probes). There is also the proposed Europa Lander, about the size of a car, which would land on the moon and investigate the surface and what lies below.

Although Europa is of intense interest, there are no actual launch dates at this time. Given the importance of and interest in Europa, it seems likely that plans will be made in the near future.

The Jupiter Icy Moons Orbiter

NASA is currently working on a project called Prometheus. The purpose of the program is to investigate nuclear-powered spacecraft. Such spacecraft would have increased power, which would allow for faster travel to and longer-duration exploration of faraway celestial bodies. It could open up a new chapter of space exploration by making previously impossible voyages possible.

As the flagship mission of the program, NASA would launch the Jupiter Icy Moons Orbiter. The orbiter would target three of Jupiter's moons: Europa, Ganymede, and Callisto. The goals of the mission

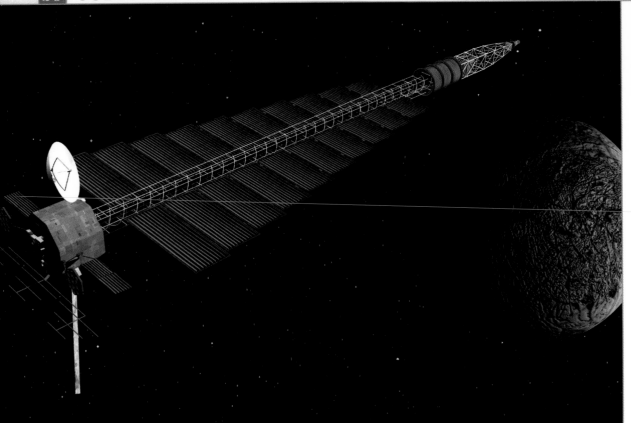

Shown here is an artist's rendition of the Jupiter Icy Moons Orbiter. The orbiter, scheduled to launch after 2011, would visit Jupiter's moons Europa, Ganymede, and Callisto. The proposed design of the orbiter would place many heat-shedding radiator panels between the spacecraft's power source and its thrusters.

would be to study the origin and evolution of the moons, the radiation environments of the moons, and the moons' potential for sustaining life. At the moment, no plans have been made for it to launch any earlier than 2011.

Is There Life Out There?

One of the reasons Jupiter's moons interest scientists so much is the possibility of finding some sort of life. Water, energy, and some sort of organic matter all lend themselves to the possibility of life as we know it. But what about life as we don't know it? Some have proposed the theory that some sort of life exists on Jupiter, a simple being that merely floats through the atmosphere.

Scientist Carl Sagan outlined such a theory in his book *Cosmos*. Sagan and his colleague E. E. Salpeter described the type of life-forms that could exist in an environment such as Jupiter's. They imagined a world of creatures existing in Jupiter's atmosphere that could propel themselves with jets of gas. Sagan and Salpeter were not proposing that such creatures actually exist, only that it is possible for such creatures to exist, that "physics and chemistry permit such lifeforms." Their model for life in such an environment shows how scientists need to use their imaginations and expand their understanding of what types of life could exist in the universe.

And what about beyond Jupiter? Some would say that the possibility of finding intelligent life is high, considering the billions of planets orbiting around billions of stars. Others disagree, believing that Earth is unique and that we are alone in the universe. One thing is certain, however. The search for life in the universe will continue. As Carl Sagan writes in *Cosmos*: "When we say the search for life elsewhere is important, we are not guaranteeing that it will be easy to find—only that it is very much worth seeking." What do you think?

1610: Galileo discovers four moons of Jupiter, later known as the Galilean moons.

1973: After passing safely through the asteroid belt, *Pioneer 10* reaches Jupiter in December. *Pioneer 10* takes the first close-up images of Jupiter and its moons before heading into the outer solar system.

1973: *Pioneer 11* is launched by NASA in April. *Pioneer 11* comes within 27,000 miles (43,452 km) of Jupiter's cloud tops and gathers information about the planet and its satellites, its magnetosphere, and solar wind. It then uses the gravity of Jupiter to adjust its course and head to Saturn.

1979: *Voyager 2* reaches Jupiter in July and eventually completes what is called the Grand Tour: a visit to Jupiter, Saturn, Uranus, and Neptune.

2003: *Pioneer 10* sends its final signal in January. The probe will continue to travel silently through space and will pass near the first object in its path, the star Aldebaran in the constellation Taurus, in 2 million years. *Voyager 1* reaches 8.4 billion miles (13.5 billion km) from the Sun on November 5.

1972: *Pioneer 10* is launched by NASA in March. Its mission is to photograph Jupiter and its satellites.

1977: *Voyager 1* and *Voyager 2* are launched in August and September to investigate Jupiter's atmosphere and magnetosphere, as well as the satellites of both Jupiter and Saturn, the ring system of Saturn, and to search for a possible ring system around Jupiter.

1979: *Voyager 1* reaches Jupiter in March and discovers that Jupiter does indeed have a ring system.

1994: In July, *Galileo* sends back images of Shoemaker-Levy 9 impacting Jupiter's surface.

2011: Tentative plans for NASA to launch the Jupiter Icy Moons Orbiter. The orbiter will study three of Jupiter's moons: Europa, Ganymede, and Callisto. The goals of the mission are to study the origin and evolution of the moons, their radiation environments, and their potential for sustaining life.

Glossary

asteroid A rocky celestial body that can measure in size from a few feet to hundreds of miles across. Most asteroids in the solar system can be found in the asteroid belt, a region between the orbits of Mars and Jupiter.

atmosphere The outer layer of gases of a celestial body.

comet A celestial ball of ice, originating beyond Pluto's orbit. When a comet approaches the Sun, it begins to melt, creating a tail.

convection An atmospheric current created by variations of density and temperature.

crust The rocky outer layer of a celestial body.

density The mass of a substance per unit of volume.

gas giant A planet made completely, or almost completely, of gas and liquid. The largest planets in the solar system—Jupiter, Saturn, Uranus, and Neptune—are all gas giants.

gravity An invisible force that causes all matter to be attracted to all other matter.

gravity assist The use of a planet's gravitational field to gain energy and speed up a spacecraft.

Hubble Space Telescope An orbiting observatory launched in 1990 in a joint project of NASA and the European Space Agency. It orbits 380 miles (612 km) above Earth.

magnetic field The portion of space near a celestial body where magnetic forces can be detected. A magnetic force is an invisible force of attraction or repulsion between objects, especially those made of iron and certain other metals.

magnetosphere The region of space surrounding a celestial body that is dominated by the object's magnetic field and that traps charged particles from solar wind.

mantle The layer of a rocky celestial object, such as a planet or moon, that lies between the crust and the core.

mass The amount of material that an object contains.

radiation Energy radiated in the form of waves or particles; high-energy particles found naturally in space in the form of solar wind particles and cosmic rays.

silicate Any of a number of mineral compounds that contain silicon, oxygen, and one or more metals. Most of Earth's crust is made of rock containing silicate minerals.

solar wind The stream of gas particles flowing away from the Sun.

terrestrial planet A planet composed of rock and metals. Mercury, Venus, Earth, and Mars are all terrestrial planets.

volume The amount of space occupied by an object.

For More Information

Ames Research Center
Moffet Field, CA 94035
(650) 604-5000
Web site: http://www.arc.nasa.gov

Goddard Space Flight Center
Code 130
Office of Public Affairs
Greenbelt, MD 20771
(301) 286-8955
e-mail: gsfcpao@pop100.gsfc.nasa.gov
Web site: http://www.gsfc.nasa.gov

Jet Propulsion Laboratory
4800 Oak Grove Drive
Pasadena, CA 91109
(818) 354-4321
Web site: http://www.jpl.nasa.gov

Smithsonian National Air and Space Museum
Seventh and Independence Avenue SW
Washington, DC 20560
(202) 357-2700
Web site: http://www.nasm.si.edu

Web Sites

Due to the changing nature of Internet links, the Rosen Publishing Group, Inc., has developed an online list of Web sites related to the subject of this book. This site is updated regularly. Please use this link to access the list:

http://www.rosenlinks.com/lnp/jupi

For Further Reading

Couper, Heather, and Nigel Henbest. *Space Encyclopedia*. New York: Dorling Kindersley, 1999.

Flaum, Eric. *The Planets: A Journey into Space*. New York: Crescent Books, 1988.

Fradin, Dennis Brindell. *The Planet Hunters: The Search for Other Worlds*. New York: Simon & Schuster, 1997.

Kerrod, Robin. *Jupiter* (Planet Library). Minneapolis, MN: Lerner Publications, 2000.

Landau, Elaine. *Jupiter*. Danbury, CT: Franklin Watts, 1999.

Lippincott, Kristen. *Astronomy* (Eyewitness Books). New York: Dorling Kindersley, 2000.

Miller, Ron. *Jupiter* (Worlds Beyond). Brookfield, CT: Twenty-first Century Books, 2002.

Spangenburg, Ray, and Kit Moser. *A Look at Moons*. Danbury, CT: Franklin Watts, 2000.

Stott, Carole. *Astronomy*. Boston: Kingfisher, 2003.

Bibliography

Institute for Astronomy. "New Satellites of Jupiter Discovered in 2003." May 28, 2003. Retrieved September 2003 (http://www.ifa.hawaii.edu/~sheppard/satellites/jup2003.html).

Johns Hopkins University Applied Physics Laboratory Press Release. "Johns Hopkins Applied Physics Lab Researchers Discover Massive Gas Cloud Around Jupiter." February 27, 2003. Retrieved September 2003 (http://www.jhuapl/newscenter/pressrelease/2003/030227.htm).

Kerrod, Robin. *Jupiter* (Planet Library). Minneapolis, MN: Lerner Publications, 2000.

NASA. Pioneer Project Home Page. Retrieved September 2003 (http://spaceprojects.arc.nasa.gov/Space_Projects/pioneer/pnhome.html).

NASA News Release. "Pioneer 10 Spacecraft Sends Last Signal." February 25, 2003. Retrieved September 2003 (http://amesnews.arc.nasa.gov/releases/2003/03_25HQ.html).

NASA. "Project Prometheus: Jupiter Icy Moons Orbiter Fact Sheet." February 2003. Retrieved September 2003 (http://spacescience.nasa.gov/missions/JIMO.pdf).

NASA. Science@NASA. "The Great Dark Spot." March 12, 2003. Retrieved September 2003 (http://science.nasa.gov/headlines/y2003/12mar_darkspot.htm?from=astrowire).

NASA. Solar System Exploration. Various pages. Retrieved August and September 2003 (http://solarsystem.nasa.gov/index.cfm).

NASA. Space Educator's Handbook. "Jupiter." Retrieved September 2003 (http://www.jsc.nasa.gov/er/seh/jupiter.html).

NASA/JPL. "Comet Shoemaker-Levy 9: The Great Comet Crash of 1994." Retrieved September 2003 (http://stardust.jpl.nasa.gov/comets/sl9.html).

NASA/JPL. "Galileo Amazing Facts." Retrieved September 2003 (http://www.jpl.nasa.gov/galileo/fact).

NASA/JPL Press Release. "Rising Storms Revise Story of Jupiter's Stripes." March 6, 2003. Retrieved September 2003 (http://saturn.jpl.nasa.gov/news/press-releases-03/20030306-pr-a.cfm).

NASA/NSSDC. "Galileo Project Information." May 22, 2003. Retrieved May 2003 (http://nssdc.gsfc.nasa.gov/planetary/galileo.html).

NASA/NSSDC. "Voyager Project Information." May 20, 2003. Retrieved September 2003 (http://nssdc.gsfc.nasa.gov/planetary/voyager.html).

NASA Astrobiology Institute. "Focus on Europa." April 13, 2001. Retrieved September 2003 (http://nai.arc.nasa.gov/news_stories/news_detail.cfm?ID=172).

Sagan, Carl. *Cosmos*. New York: Ballantine Books, 1985.

Space.com. "What Is a Moon? Definition Lags Behind Soaring Satellite Tally." April 1, 2003. Retrieved September 2003 (http://www.space.com/scienceastronomy/moon_definition_040103.html).

Watters, Thomas R. *Planets* (Smithsonian Guides). New York: Macmillan, 1995.

Index

A
Adrastea, 32
Amalthea, 23, 32
asteroid belt, 7, 9, 12
Atlantis, 12
auroras, 30

C
Callisto, 19–20, 35
Cassini, 12, 24, 27, 28, 30
convection, 26

D
Del Genio, Dr. Tony, 28
Discovery, 12

E
Earth, 5, 10, 12, 13, 14, 15, 16, 17, 21, 22,23, 25, 26, 27, 28, 30, 37
Europa, 6, 13, 22, 32, 33–35
Europa Focus Group, 33, 35
Europa Ice Clipper, 35
Europa Lander, 35
Europa Ocean Explorer, 35
European Space Agency, 10–11

G
Galileo, 6, 12–13, 19, 20, 21, 25, 31
Galileo Galilei, 5, 6, 18, 20
Ganymede, 18–19, 35
gas giants, 14, 30, 31
Grand Tour, 10
gravity assist, explanation of, 11–12
Great Dark Spot, 28

Great Red Spot, 28
Greeley, Ron, 33–35

H
Himalia, 22–23
Hooke, Robert, 28
Hubble Space Telescope, 30

I
International Astronomical Union, 20
Io, 12, 20–21, 32
Io plasma torus, 32
Italian Space Agency, 12

J
Jewitt, Dr. David, 22, 24
Jupiter
    atmosphere and composition of, 6, 7, 9, 13, 25–30, 33, 37
    density of, 15
    diameter of, 14
    gravitational pull of, 17, 22
    magnetosphere of, 7, 9, 12, 30, 31–32
    mass of, 6, 14–15, 17
    moons of, 5, 6, 7, 9, 10, 12, 13, 17, 18–24, 32, 33–36
    radiation/radiation belts of, 6, 7, 12, 25, 30, 32, 33
    ring system of, 9, 10, 12, 30–32
    rotation of, 15
    seasons on, 28
Jupiter Icy Moons Orbiter, 35–36

K
Kepler, Johannes, 20

M

magnetosphere, explanation of, 7, 30

Marius, Simon, 20

Mars, 5, 6, 7, 14, 16, 19

Mercury, 14, 18, 19

Metis, 32

N

NASA, 7, 10, 12, 13, 23, 27, 28, 35

Neptune, 10, 14

P

Pioneer spacecraft/missions, 6, 7–9, 10

Pluto, 14, 18

Prometheus, 35

S

Sagan, Carl, 37

Salpeter, E. E., 37

Saturn, 9, 10, 12, 14, 16, 24, 30, 31, 32

Shoemaker-Levy 9, 13

solar wind, 9, 28, 30

T

terrestrial planets, 14

Thebe, 32

tidal forces, explanation of, 22

Trojans, 17

U

Ulysses, 10–12

Uranus, 10, 14

V

Valhalla, 20

Venus, 5, 12, 14

Voyager spacecraft/missions, 6, 9–10, 18, 30, 31

About the Author

Abby Young is a freelance writer living in Brooklyn, New York.

Photo Credits

Cover, pp. 12, 19, 21, 23, 26, 29, 31 (left), 34, 36 NASA/JPL/CalTech; pp. 4–5 NASA/JPL/Space Science Institute; p. 8 NASA/Ames Research Center; p. 11 NASA/Kennedy Space Center; p. 15 © Réunion des Musées Nationaux/Art Resource, NY; p. 16 © Rev. Ronald Royer/Science Photo Library; p. 27 NASA/JPL/ University of Arizona; p. 31 (right) NASA/JPL/Johns Hopkins University Applied Physics Laboratory.

Designer: Thomas Forget; Editor: Nicholas Croce